BLUE-G

THORSONS NUTRIENTS FOR HEALTH

BLUE-GREEN ALGAE

—

PAUL SMITH, PH.D.

Thorsons

An Imprint of HarperCollins*Publishers*

Thorsons
An Imprint of HarperCollins*Publishers*
77–85 Fulham Palace Road
Hammersmith, London W6 8JB

1160 Battery Street
San Francisco, California 94111–1213

Published by Thorsons 1996
1 3 5 7 9 10 8 6 4 2

A catalogue record for this book
is available from the British Library

ISBN 0 7225 3319 5

Printed in Great Britain by
Caledonian Book Manufacturing Ltd, Glasgow

For Tippawan

The information in this book is advisory in nature and does not form a substitute for professional medical treatment. If in doubt, consult your GP or alternative practitioner.

Contents

Introduction

Green and blue-green algae have been harvested for hundreds of years, and the twentieth century has seen them increasingly valued as a rich source of food and food supplements.

As both a food and a chemical resource, algae have become a significant part of the local economy in many parts of the world, including China, India, Japan, South America, Taiwan, Thailand and the United States. The health-food-quality algae industry is said to be worth over $100 million a year and growing, and anecdotal claims for the powers of algae are currently catching the attention of the world's media, especially since the dried green powder has been endorsed by the famous.

One article in a serious English Sunday newspaper last year claimed that blue-green algae can 'boost vitality ... improve sleep, reduce allergies, stop migraines, reduce stress, alleviate PMT and boost the body's immune system' – far-reaching claims indeed. This is the gossip – gossip that has included the names of Demi Moore, Jack Nicholson, Kate Moss and Björk. Madonna is said to have made a point of taking it before getting pregnant, and you are more likely to see young people in any metropolis sipping 'smoothies' containing algae than drinking champagne.

But is all this just hype? In this health guide to blue-green algae I have looked beyond the pages of the press – popular and serious – and the publicity material sent out by the health food industry, and I have examined the scientific, medical and pharmaceutical evidence. And it is clear from reading the results of well-documented research that there are secrets to be unlocked in the many species of algae that inhabit the planet.

There does indeed seem to be evidence that algae can help with a variety of ills. Clinical tests are at present being conducted around the world to confirm the beneficial effects of algae on a bewilderingly diverse list of medical conditions, and positive research findings are fast appearing in the scientific literature.

I have tried here to give a brief but accessible, readable and authoritative general introduction to current knowledge of green and blue-green algae, and to provide nutritional analyses and dietary information in an easy-to-understand way. I take a critical look at the medical research findings and the health claims made for microalgae and their derivatives. I also report news of current research into the commercial exploitation, by pharmaceutical and other industries, of this resource which, we are told, is going to save the world.

1

What are Blue-green Algae?

There are many hundreds of different algae throughout the world. They range from the various types of seaweed that you find washed up on the beach to the scummy green stuff that gathers on stagnant ponds and in swimming pools. Algae range in size from giant kelp, that can grow up to 50 metres in length, to varieties that are no more than tiny individual cells which are so small they have to be measured in 'microns' or 'micro-metres', and there are one million micro-metres to the metre. The smallest ones are called microalgae, and the blue-green algae capsules or tablets that you buy in health food shops contain these tiny single-cell or multicellular plants that turn sunlight into life energy.

Not all blue-green algae are actually blue-green in colour. Some are more of a green-blue, and you can find red, yellow, black or brown varieties. The Red Sea actually looks red because it is full of tiny reddish algae.

You can find algae virtually anywhere, and many of them, like a few types of seaweed and some of the kinds that crop up in fresh-water lakes, have been used as food for centuries. Once they have established themselves algae seem to keep on spreading.

In the last 40 years several green and blue-green algae

have been singled out for their nutritional and chemical properties, and a great deal of research world-wide is concentrated on the mass cultivation of microalgae, their many properties and the exploitation of their pharmaceutical and industrial potential. Seaweed is a delicacy in many parts of the world – in Wales, for example, as well as in America and the Far East – and it is a component of the staple diet in some areas of Africa.

For many years, especially in Japan, green algae have been valued as a nutritious food, and they have been eaten both as an everyday ingredient in the diet and as a health food supplement. Specific species of green and blue-green algae have been promoted by producers, and some varieties of dried algae are marketed under brand names. The retail cost of microalgae supplements that are sold loose, as tablets or in capsules varies, but they can command relatively high prices in Japan, the United States and now increasingly in Europe.

Much has been written, to begin with in the health food press, but now more frequently in popular newspapers and magazines, about the health claims made by manufacturers and distributors of green and blue-green algae. They have been described as a 'superfood', and the word 'probiotic' figures largely in the marketing literature. Movie stars and television personalities have endorsed blue-green algae, making claims for their energy-enhancing and rejuvenating properties, and the media have picked up on reports in scientific journals of medical and pharmaceutical breakthroughs based on research into microalgae and cyanobacteria (which is the true name for blue-green algae).

Most commercial attention has been focused on the true algae Chlorella, Dunaliella and Scenedesmus, and the cyanobacteria Aphanizomenon flos-aquae and Spirulina. This is because they are the most easily cultivated artificially, but there are thousands more species than these. Microalgae are extremely adaptable and are found in most environments – from dry sun-drenched deserts to hot springs and icy lakes – and they grow at a phenomenal rate, often richly colouring ponds and lakes. Because of their ability to thrive in what are often extreme conditions, green and blue-green algae have been nicknamed 'the creatures of earth, wind and fire'.

Single-celled microalgae are extremely small (you need a microscope to see one), and in one square centimetre of water in a lake there could be 100 million of them. Their ability to reproduce is breathtaking – a single Chlorella can increase 40 times its mass in a single day. Every year one acre can produce 1,000 tons of algae that dries to 30 tons of powder. Nutritionally Chlorella is extremely rich. Almost 70 per cent of its dry weight is protein and it has been estimated that one acre of land growing algae will yield 10 tons of protein compared with 0.16 tons of wheat. So, understandably, great things have been predicted for microalgae, with claims that they will end world famine and make developing countries self-sufficient.

There are thus two strands to the 'microalgae as super-food' question. First, could microalgae feed the hungry – providing cheap and effective nutrition for the world's malnourished poor? Second, if microalgae as a food supplement has life-enhancing and perhaps life-saving properties, can it provide increased energy, protection against disease and other health benefits if consumed alongside a traditional diet?

2

Eating the Stuff

Microalgae are attractive to nutritionists for many reasons. They are found in, or can be successfully introduced into, parts of the world that would be considered hostile to conventional agriculture – areas with excessive heat and sunlight or with available water ranging from salty seas to alkaline and high-salt lakes and waters. Microalgal cultivation requires little or no artificial herbicides or pesticides, and is energy efficient. It does not erode, deteriorate or pollute the natural existing environment, and in the absence of the right conditions artificial cultivation is easily undertaken. Indeed, many millions of dollars of aid from the United Nations and the World Bank have been spent on schemes to develop the cultivation of microalgae around the world.

It would be reassuring to think that at least some of the research into microalgae could benefit developing countries and help even out the balance of food and energy resources world-wide. About 10 per cent of the world's population is undernourished.

Both Spirulina and Scenedesmus have been used in countries where malnutrition is common. In trials in Peru, badly malnourished children were given 10g of Scenedesmus, and babies who were nearly dead from malnutrition were given around 1g per kilo of body weight. The results were dramatic, proving that algae can actually improve the absorption of food by activating the regeneration of intestinal epithelium. In parts of India the blue-green alga Spirulina has been used as a basic foodstuff for undernourished children, and has proved more successful than soya, which until recently has been the main nutrient-rich food supplement supplied by aid agencies.

As plants, complete in themselves, microalgae are incredibly rich, containing a balance of nutrients that make them virtually a 'whole food' – capable of sustaining life without the need for other foods. Given the crucial criteria of cost, land use and energy efficiency, microalgae would seem to be the perfect answer to those seeking a cheap, reliable nutritious diet. They grow in climatic extremes of heat and cold, in poor water or in no water at all and can be cultivated with a minimum of expertise.

Certainly a breakdown in nutritional terms of a few of the most commonly available microalgae reveals an impressive list of nutrients and the distributors of microalgae make outstanding claims for their products. But what are nutrients and what exactly does the fine green powder in an average capsule of Aphanizomenon flos-aquae or Chlorella actually provide in the way of nutrition?

The Body's Need for Nutrients

Nutrients are substances that the body requires in the diet for growth, maintenance of health and reproduction. Protein, carbohydrates and fat provide energy, and vitamins and minerals are required for other substances in the body to carry out their functions. Proteins also have other roles, including catalysing chemical reactions, transporting substances around the body in the blood, maintaining fluid balances and acid-base balances and controlling muscle actions. Nutrients interact in complex ways: for example, vitamin C and copper interact to allow iron to function properly and each is needed in the correct proportions to regulate a series of bodily functions; zinc in different quantities affects the absorption and utilization of copper.

For most people in the developed world a varied diet will contain all the vitamins and almost certainly the vast majority of the minerals that are required to stay healthy. In developing countries diet can be insufficient in vitamins and minerals, and in children this can have serious effects on growth rates, intellectual development and general health. Deficiency of a single nutrient can impair body processes, and deficiency over a long period of time can lead to death.

In the developed world, however, multivitamin and mineral supplements are big business. (One survey in 1987 estimated the health food market in Chlorella and Spirulina alone to be worth some $100 million worldwide.) In exceptional circumstances it may be necessary to add to your diet foods rich in certain vitamins or minerals, and your GP will know when a specific medical condition demands a vitamin or mineral supplement, and will recommend the necessary changes in diet.

There are many misconceptions regarding diet and, in particular, the use of vitamin and mineral supplements. Taking vitamin supplements cannot boost your energy. If you are diagnosed as being deficient in a vitamin and you start eating the correct foods or taking a vitamin supplement you will not see any instant effects. It might take weeks or months for the imbalance or deficiency to be rectified and for any benefits to become apparent. Vitamins are not going to increase intelligence or improve your memory. Reports that taking multivitamins can give children an advantage at school have been discredited. Nevertheless they have become firmly established in dietary folk-myths which newspapers and magazines tend to pick up on and repeat.

There are various circumstances under which vitamin and mineral supplements may be needed. If you follow a strict vegetarian diet or are on a diet to reduce weight, if you are taking the contraceptive pill or if you are pregnant, or planning to become pregnant, then it may be advisable to add certain foods to your daily diet or to take multivitamins.

Although it is perfectly possible to eat a well-balanced diet as a vegetarian it is not always so easy. And if you are a strict vegan and eat only food from plant sources you must make sure that you include enough calcium, riboflavin, iron, zinc, vitamin B_{12} and vitamin D in your diet, many of which non-vegetarians will obtain by drinking milk and eating milk products, eggs and meat. Care must also be taken to combine proteins from plant sources so as to provide all the essential amino acids. A varied amalgamation of green leafy vegetables, legumes, fruits, nuts and grains can rule out any deficiency, exposure to sunlight can supplement vitamin D and only vitamin B_{12} will need to be taken as a supplement.

Spirulina has the highest vitamin B_{12} content of any unprocessed plant or animal food and so is an ideal source of the vitamin for vegetarians. It was generally believed before the analysis of the microalgae that vitamin B_{12} was available in substantial amounts only from foods derived from animals. Another, more restrictive, vegetarian diet is the Zen macrobiotic diet which at its highest level allows only cereal grains – by definition ruling out the taking of any supplement – and this regimen can lead to life-threatening malnutrition.

Taking the contraceptive pill can reduce the body's absorption of vitamins B_6 and B_{12}, vitamin C and folic acid, and, interestingly, increase absorption of vitamins A and K. Some nutritionists recommend that you should take this into account if you are taking the oral contraceptive, especially if you are underweight or dieting. The lower vitamin B_6 intake may account for the depression that some women suffer from while taking the contraceptive pill. A number of companies produce multivitamin supplements specifically for women who are taking the pill, but foods rich in these nutrients will in all likelihood form part of most women's diets (eggs, meat, offal, vegetables, wholegrain cereals for vitamin B_6; red meat, dairy products, eggs, milk, offal, seafood for vitamin B_{12}; citrus fruits, green vegetables, peppers, potatoes, tomatoes for vitamin C; lambs' livers, leafy green vegetables, pigs' livers, soya flour, wheatbran, wheatgerm for folic acid).

Pregnancy and breast-feeding are important factors in terms of nutritional requirements. A woman's diet both before and during pregnancy has a direct effect on the birthweight of her baby. A woman and her partner who are trying to conceive are recommended to increase their intake of the B-vitamins and folic acid, especially if she has recently been taking the contraceptive pill (see above). Deficiencies in folic acid have been associated with an increased risk of having a child with a neural tube defect such as spina bifida. A deficiency in iodine can lead to mental and physical retardation. Babies born to anaemic women are more likely to be of low weight and the incidence of infant death can be higher.

3

The Nutritional Value of Algae

Edible algae are principally composed of protein, carbohydrates (sugar or vegetable gums), fats and ash (sodium and potassium). They also contain, in varying amounts, most of the vitamins and minerals.

The following table shows up-to-date analyses of the cyanobacteria Aphanizomenon flos-aquae and Spirulina, and the true alga Chlorella, which at present are the microalgae most readily available commercially as food supplements. It should be noted that nutritional tables can only offer approximate values, and consequently they are guides rather than scientific comparisons. In the first place the individual materials should ideally be analysed at the same time, using the same methods, and by the same laboratory; second, no plant material can be expected to be uniform in its nutrients since so many factors have to be taken into account, like the medium in which the plants are grown, cultivation and harvesting techniques, the age at which the plants are harvested and the time taken to process the plant materials.

Also, the procedures used in drying microalgae can affect the concentrations of nutrients: some vitamins (e.g. nicotinic acid, ascorbic acid and vitamins B_1 and B_2) are heat-unstable. Others deteriorate with time. Two handfuls of carrots from the same crop will not produce identical readings when their nutritional content is analysed. And different crops of carrots may be even more variable.

Nutritional values of commercially available microalgae per 10g dry weight

	Spirulina	Chlorella	Aphanizomenon flos-aquae
Water	7%	5%	6%
Protein	71%	58%	63%
Carbohydrate	18%	23%	25%
Lipids	5%	9%	5%
Glycolipids	200mg	n/a	n/a
Sulpholipids	10mg	n/a	n/a
Fibre	0.9%	n/a	n/a
Ash	9%	3%	7%
Vitamins			
Beta-carotene	2300 RE*	550 RE	2000 RE
Thiamine	0.31mg	0.17mg	0.03mg
Riboflavin	n/a	n/a	570g
Niacin	1.46mg	2.38mg	0.65mg
Vitamin B$_6$ (Pyridoxine)	80μg	140μg	67μg
Pantothenic acid	10μg	130μg	130μg
Biotin	0.5μg	19μg	3.6μg
Folic acid	1μg	2.7μg	1μg
Vitamin C (Ascorbic acid)	0.5mg	1mg	5mg
Choline	n/a	n/a	2.6mg
Inositol	6.4mg	13.2mg	n/a

Vitamin B$_{12}$			
(Cobalamin)	3.2µg	1.3µg	8µg
Vitamin E	1 IU	0.1 IU	1.2 IU

Minerals

Calcium	100mg	22mg	140mg
Chlorine	n/a	n/a	46mg
Chromium	28µg	n/a	40mg
Copper	120µg	10µg	60µg
Iron	15mg	13mg	6.4mg
Magnesium	40mg	32mg	16mg
Manganese	0.5mg	n/a	0.3mg
Molybdenum	n/a	n/a	33µg
Phosphorus	90mg	90mg	51mg
Potassium	120mg	90mg	100mg
Selenium	n/a	n/a	6.7µg
Sodium	60mg	n/a	38mg
Zinc	0.3mg	7mg	187µg

Other Compounds

Nucleic acids	4.5%	30%	n/a
Gamma-linolenic acid	135mg	n/a	n/a
Carotenoids	37mg	n/a	n/a
Phycocyanin	1500mg	n/a	n/a
Chlorophyll	115mg	200µg	300mg

IU = International Unit; mg = thousandth of a gram; µg = millionth of a gram.

RE = retinol equivalent. One RE is defined as 1µg of retinol or 6µg of beta-carotene, or 12µg of other vitamin A carotenoids. A Recommended Dietary Allowance (RDA) in terms of retinol equivalents is 1000 RE per day for adult males and 800 RE per day for adult females.

The above information was provided by Cell Tech, Oregon, USA, Sun Chlorella, Japan and Sunstream, London, UK.

Charts and tables similar to these can look impressive, and often they are used by manufacturers and distributors in publicity and promotional material. But what exactly do all these figures mean and how do our bodies use the component nutrients?

Protein

Proteins are complex molecules consisting of chains of amino acids formed from a combination of carbon, hydrogen, nitrogen, oxygen and sulphur. The body requires proteins for a vast array of chemical reactions that, among other things, regulate metabolism and maintain health. Nearly all foods contain some protein. In the developed world the most important sources of protein are dairy products, eggs, fish, red meat, poultry, milk, grains, nuts and seeds, peas and beans (legumes). It is the quality of the protein that is important rather than simply the quantity consumed and quality depends on the amounts of amino acids contained in a protein. The more closely the protein matches the body's requirements for essential amino acids the higher its quality.

Twenty-two amino acids have been identified, some of which the body can produce itself (non-essential amino acids), and some of which the body needs to ingest from foodstuffs (essential amino acids). The body is constantly breaking down and synthesizing proteins to ensure correct balance and to provide energy.

All the amino acids have to be present for amino acid break-down and rebuilding to take place, and a deficiency of a single essential amino acid can interfere with your body's ability to repair itself and to grow. The essential amino acids are: histidine (essential for sustained growth in children); isoleucine; leucine; lysine; methionine (combines with cysteine, a non-essential amino acid: a diet deficient in methionine can effect the correct functioning of the liver); phenylalanine (combines with tyrosine, a non-essential amino acid: essential for psychological health); threonine; tryptophan; valine.

All the essential amino acids are found in microalgae, and as such they can be described as providing a 'complete protein'. Indeed, the amino acid composition of Spirulina protein is probably the optimum for plant materials, better even than that found in legumes and soybean products.

Carbohydrates

Carbohydrates are chemical compounds containing carbon, hydrogen and oxygen and are classified as monosaccharides and disaccharides (the sugars) and polysaccharides (starches and fibre). The carbohydrates in microalgae are starches and fibre, in the form of cellulose. Carbohydrates are a source of calories that provide energy and which assist in body processes such as the activity of nerves, muscle contraction, heart beat and breathing. Carbohydrates also combine with proteins and other nitrogen-containing compounds to give glycoproteins which make up enzymes, hormones, skin and bones. Microalgae are, on average, one-fifth starch and, as plant material, consist of an appreciable amount of fibre. The cellulose in microalgae cannot be digested by humans, since we do not have the enzymes needed to break down the cell walls.

Lipids

Fats and cholesterol are forms of non-water-soluble chemical compounds called lipids. Vitamins A, D, E and K are fat-soluble vitamins and we need to include fats in our diet so that these vitamins can be absorbed in the digestive tract. Fats from plant materials are considered generally healthier than fats from animal sources.

Microalgae also contain the polyunsaturated fatty acids linoleic and gamma-linolenic acid (in a form said to be inferior only to milk). Linoleic acid is an essential fatty acid, which means that to remain healthy we must eat a certain amount every day since the body cannot make its own supply. Linolenic acid promotes healthy growth.

Dietary lipids from marine unicellular algae are said to enhance the amount of liver and blood omega–3 fatty acids in rats. A fatty acid is derived from fats in the diet. Omega–3 is a fatty acid reportedly playing a role in lowering cholesterol and low-density lipoproteins. Diets containing algal material have been found significantly to reduce the relative abundance of arachidonic acid in the blood and liver lipids, and to cause a significant increase in the percentage of the omega–3 polyunsaturated fatty acids. The effects on rats may not sound of great interest to non-rats but there has recently been research into feeding hens microalgae and thereby increasing the levels of omega–3 fatty acids in eggs.

Fibre

Fibre has increasingly been seen as important in the diet, with certain types of fibre being able to lower blood cholesterol concentrations. Although human digestive enzymes cannot break the wall of cellulose – the cell walls of plants – bacteria in the large intestine can to a certain extent ferment it.

Ash

The percentage of ash in a nutrition table refers to the mineral matter that is left after foods are burnt in oxygen. Other components of foods – the carbohydrates, protein and fats – are removed as gases. In the body ash is excreted in the urine.

Vitamins

Vitamins are sometimes described as organic in that, unlike minerals, they contain carbon. A few vitamins the body can manufacture for itself: niacin, vitamin A and vitamin D can be synthesised provided that the necessary 'provitamins' – the raw materials – are present in the diet. Vitamins are either fat-soluble or water-soluble (see above). Fat-soluble vitamins (A, D, E and K) can be stored up in the liver and so do not need to be present in the diet on a daily basis. Water-soluble vitamins, on the other hand, are not stored in the body – on the contrary, they are constantly being excreted in urine. These vitamins are required to be eaten every day. Moreover, they deteriorate in foodstuffs and are depleted during storage and cooking.

Microalgae are rich in the following vitamins:

Beta-carotene

An antioxidant, beta-carotene (*cis* and *trans*) is converted in the liver into vitamin A, which is essential for the maintenance of mucus membranes, in particular the respiratory tract, and for fertility and growth. Beta-carotene is an important substance for the correct working of the immune system and for the maintenance and repair of nails, hair, teeth and bones. A lack of beta-carotene in the diet can make you more prone to common infections and cause certain skin conditions. Beta-carotene and its derivatives are at present being tested as chemopreventives for certain cancers. Research at the University of Buffalo in the US on the diet of men who smoke has shown that eating food rich in beta-carotene can reduce the likelihood of their developing cancer of the prostate.

Vitamin B$_1$ (thiamine)

Thiamine is a coenzyme required by the body for the correct functioning of the nervous system. Thiamine deficiency can impair nerve/tissue function and affect the digestion, the heart and proper growth.

Vitamin B$_2$ (riboflavin)

Riboflavin is a coenzyme necessary for correct metabolism. It is involved in oxidation-reduction. A lack of riboflavin can lead to skin conditions, eyesight problems, a swollen tongue and cracked and sore lips.

Niacin

A coenzymexe involved in oxidation-reduction reactions in metabolism, niacin is a B-vitamin necessary in the diet for overall health. Niacin deficiency can lead to mental confusion, ulcers in the mouth and digestive disorders, including diarrhoea.

Vitamin B$_6$ (pyridoxine, pyridoxal and pyridoxamine)

Pyridoxine, pyridoxal and pyridoxamine are coenzymes used by the body in the assimilation of protein in the diet. A deficiency can affect growth and vital bodily functions and impair the immune system. Symptoms include anaemia, nausea, dizziness, weight loss and skin conditions.

Pantothenic acid

A coenzyme that aids the release of energy from carbohydrates, fats and proteins, pantothenic acid is essential for balanced hormone production and proper growth. Symptoms of deficiency (which is extremely rare in humans) include fatigue, nausea, headaches and muscle debilitation.

Biotin

A coenzyme involved in the synthesis of glucose, fatty acids and nucleic acids. Deficiency symptoms are depression and apathy, loss of appetite, hallucinations and flaky skin.

Folic acid

A coenzyme involved in the synthesis of nucleic acids and the metabolism of amino acids and choline, of particular importance for women who are pregnant or who are trying to get pregnant. Deficiency can lead to anaemia and various skin conditions.

Vitamin C (ascorbic acid)

Ascorbic acid is an antioxidant and is essential for a normal healthy metabolism. A deficiency of vitamin C leads to scurvy, and signs can be bruising easily, bleeding, weight loss, a general lack of energy and irritability. Ascorbic acid has long been thought to have therapeutic properties, in particular for the functioning of the immune system if taken in larger quantities than are absolutely necessary, and research is currently being carried out to test these claims. Some people have promoted very high doses of vitamin C as a cure for the common cold, but there are dangers of overdosing which can lead to problems with the digestive tract. Vitamin C is being investigated in Italy and the United States as a cancer chemopreventive.

Choline

Part of the phospholipid lecithin, choline is designated a vitamin in its own right only by some nutritionists, since it is readily synthesized in the body. Choline is an important component of cell membranes and is involved in the movement of lipids in the blood. It is vital for the correct assimilation of fats, preventing them from overloading the liver. Choline deficiency can affect both the kidneys and the liver.

Inositol

Inositol is not really a vitamin since, like choline, the body can synthesize it and deficiency is unknown in humans. Nevertheless, the promoters of health food supplements frequently brand their products as being rich in inositol.

Vitamin B_{12} (cobalamin)

A coenzyme involved in the metabolism of amino acids, cobalamin is generally found in animal products only. Deficiency can lead to pernicious anaemia and to degeneration of the spinal cord. It is essential for the correct functioning of the nervous system and the digestion. Work in the Netherlands suggests that vitamin B_{12} derived from algae and other plant food may not be as easily assimilated as it is from other sources.

Vitamin E

An antioxidant, vitamin E prevents the deterioration of vitamin A, vitamin C and polyunsaturated fatty acids. Too much vitamin E can make you nauseous or can even stop the blood from clotting. A deficiency in children can lead to anaemia but is unlikely to affect adults. In a similar way to vitamin C, vitamin E has been promoted by 'vitamin gurus' as having extraordinary powers including enhanced athletic and sexual performance. No evidence has yet been found to substantiate such claims.

Minerals

Minerals function with or as parts of other compounds and are of particular importance in the production and working of enzymes. They are needed to maintain a significant number of bodily processes like growth, nerve and muscle function, blood system and reproduction. Some minerals (macronutrients) are needed in relatively large quantities (more than 100mg) every day. These are calcium, chloride, magnesium, phosphorus, potassium, sodium and sulphur. Other minerals (micronutrients), sometimes referred to as trace elements, are only needed in minute quantities. These are chromium, cobalt, copper, fluorine, iodine, iron, manganese, molybdenum, nickel, selenium, silicon, tin, vanadium and zinc.

    Microalgae are rich in the following minerals:

Calcium

Calcium is, of course, essential for healthy bones and teeth and is important in the normal function of blood clotting. The mineral is particularly crucial in the diets of babies and of menopausal and post-menopausal women. Calcium has been undergoing tests in Lyon in France for its effectiveness in preventing colorectal cancers in those considered at high risk. Experiments feeding increased dietary calcium to mice have led to decreased cell proliferation and decreased colon cancer.

Chlorine

Chlorine is a constituent of the fluids surrounding the cells and an adult body contains about 70g. Most of the chlorine in our diet is in the form of salt (sodium chloride). Adopting a salt-free diet can considerably reduce the intake of chlorine but deficiency is not generally a problem unless there is a corresponding lack of sodium in the diet. Chlorine is discarded from the body in sweat and urine.

Chromium

A deficiency in chromium can produce a condition similar to diabetes. It is an aid in the metabolism of glucose.

Copper

Involved in the metabolism of iron, copper is also a co-factor for a number of enzymes including those involved in myelin synthesis, pigmentation and connective tissue. Copper deficiency can be a cause of anorexia contributing to bone disease, reducing white cell count and producing anaemia.

Iron

Deficiency in iron can lead to anaemia and a general feeling of sluggishness. Iron is necessary for blood cells to transport oxygen around the body and for a healthy liver and spleen.

Magnesium

Most healthy diets are sufficient in magnesium which is essential for the correct functioning of enzymes in the body. Magnesium is necessary for proper muscle function and has been shown to be beneficial in the treatment of asthma.

Manganese

A lack of manganese can lead to nausea and weight loss combined with certain skin conditions.

Molybdenum

Molybdenum is an enzyme co-factor which is important for the transfer of the oxygen atom from water to various compounds. Deficiency is unknown in humans, but in animals it can affect growth and lead to higher mortality for both mother and offspring in the birth process.

Phosphorus

Essential for the correct absorption in the intestine of amino acids, phosphorus is a component of body materials including bones and teeth, and is required for the metabolism of glucose. Readily available in the diet from meat, dairy products, nuts and grain, deficiency is rare.

Potassium

Co-factor for enzymes and essential for proper nerve and muscle function, potassium is readily available in fruit, vegetables, nuts, meat, fish and dairy products. Deficiency is unlikely, although in severe malnutrition symptoms would include rapid heartbeat, nausea and weakness.

Selenium

A co-factor for a peroxide-neutralizing enzyme. The peroxides come from polyunsaturated fatty acids and are also produced by metabolic reactions. Selenium is an antioxidant and is being investigated as having a role in the prevention of cancers of the head and neck. In large doses selenium has been shown to have toxic effects on animals that have eaten grain from soil that has this element in higher than normal concentrations, but the only evidence of ill-effects in humans is of a greater incidence of tooth decay in children. There are only traces of selenium in microalgae.

Sodium

Most microalgae have a low sodium content and all health authorities maintain that a reduction in sodium in the form of salt is desirable in the diet.

Zinc

An antioxidant, in the last 20 years traces of zinc in the diet have been found to be essential for the speedy healing over of cuts and grazes, and a lack has been associated with a loss of the sense of taste. Zinc is not easily absorbed and so even though it is commonly found in fairly significant quantities in many foodstuffs deficiencies can occur in otherwise healthy people.

Other Components

Microalgae contain other components, including nucleic acids which are involved in complex protein synthesis sequences in the body. Carotenoid pigments are vitamin A provitamins. Phyco means seaweed – phycocyanin is the blue colouring matter which is combined with chlorophyll in certain algae. Precisely what the benefit of chlorophyll is to humans is still under examination. The health food press makes claims about chlorophyll aiding the assimilation of nutrients into the bloodstream. Certainly 'greens' should be a key part of anyone's diet, but the importance of the chlorophyll in the equation is as yet unknown.

Eating Microalgae

The mix of nutrients that microalgae contain make them appear to be the ideal food. And if you haven't ever seen, let alone tasted, dried Aphanizomenon flos-aquae, Chlorella or Spirulina you may find it difficult to appreciate that microalgae and cyanobacteria are not the most attractive foods on the planet. The thought of using microalgae as a staple food is difficult to take seriously.

A report published in the *Journal of Nutrition* at the beginning of the 1960s provides a detailed investigation into the digestibility of Chlorella and Scenedesmus. For just over one month (35 days) four men were fed dried microalgae starting with 10g per day, increasing at five-day intervals to 500g per day. Levels of 10g and 20g per day were tolerated, but greater amounts of microalgae produced intestinal discomfort and flatulence. A very few people find even the small amounts of microalgae or cyanobacteria in the tablets or capsules sold as health food supplements can produce similar effects. Moreover, it should be remembered that the intake of high amounts of single-cell products is not recommended since the nucleic acid content can become metabolized by the body into uric acid which can lead to gout.

More recent reports have discovered that digestibility is dependent on the drying techniques applied to the microalgae, the way the microalgae are cooked or prepared and the types of foodstuffs that the microalgae are combined with. Certainly it has been found that freeze-drying contributes to overall digestibility and that microalgae added to hot food is best tolerated.

Microalgae as a Health Supplement

Microalgae have been labelled 'designer foods' – a term that looks impressive in marketing handouts and in health food journalism, and the market for green and blue-green algae has increased steadily over the last 10 years. Microalgae are claimed to be easily assimilable – and in the quantities recommended by the producers of health food supplements this is true for most consumers – and they do present a remarkable mixture of essential nutrients.

There is little doubt that microalgae can provide a valuable source of supplementary vitamins and minerals. Some microalgae have been suggested as a source of additives for infant feeding formulas. Those that are rich in arachidonic acid, which is also a component of maternal human milk, could be used to enhance the nutritional value of artificial baby food. However, unlike the Far East, there is not a tradition of using seaweed and other algae in our cuisine, and the demand for microalgae, and perhaps more specifically for blue-green algae, has relied on anecdotal reports of its miraculous properties and to a large degree on fashion. It is unlikely that the health food supplement market itself will do anything but continue to expand and if the price of materials like dried microalgae ceases to be so inflated, then perhaps the attractiveness of such a balanced mix of vitamins and minerals, marketed as a 'wholefood' derived from plant material, can be maintained.

As a health supplement, of course, nowhere near the quantities described above in the *Journal of Nutrition* are recommended, nor would the relative cost of commercially available green and blue-green algae make this practicable. In the United Kingdom, Chlorella extracted from the water in which it is cultivated – in other words, which has not been dried – is being marketed added to drinks containing fruit juice. Several books from Japan and the United States have included recipes for 'smoothies' – drinks that are cocktails of fruit and vegetable juices and microalgae or blue-green algae.

Super Healthy Low-calorie
Klamath Lake Algae Smoothie Drink

Ingredients

    50cl red grape juice
    1 ripe banana
    half a pot of organic live plain yogurt
    1g Klamath Lake blue-green algae powder
    2 mineral water ice cubes

Directions

    Blend all ingredients in a liquidizer until smooth, and pour into a glass.

    Vary the fruit, adding strawberries, mangoes, kiwi fruit or oranges. This smoothie is quite simply the healthiest way to start the day.

Algae are best eaten combined with other foods – preferably with stronger flavours. In Holland, for example, a blend of tapioca, wheat flour, maize grits, the microalga Spirulina and seaweed Wakame and Nori have been used to produce a snack food to compete with crisps, tortilla chips and similar products. The food was analysed and compared with the largest-selling available equivalents, and the new snacks were found to have lower fat content and relatively higher calcium and iron. Blind consumer tests were largely favourable, although people's perceptions of exactly what are microalgae and algae were not.

Away from the health food market (where the consumer is well informed and more prepared to try the unfamiliar) it seems that there could well be a reluctance by the general consumer to try a product made from seaweed or other algae. However, algal and microalgal materials can be listed as vegetable material and so concealed in the ingredients, as in the case with the snack food.

4

Health Benefits

Many microalgae and algae, especially seaweed, have been valued for their medicinal properties for thousands of years. Chinese documents dating from 2700 BC mention marine algae as being recommended for the treatment of unspecified maladies, and the interest generated by microalgae in the last ten years or so probably derives in no small measure from the high regard in which algae of all kinds are held in Japan and other parts of the Far East.

It is a matter of statistical record that the Japanese diet is far healthier than the western diet, and a lot of work has gone into identifying precisely what the Japanese diet includes that the western diet lacks. The incidence of obesity, heart disease, diabetes and certain cancers are considerably higher in western countries than in Japan. In the west, people over the age of 50 – particularly men – have a greater likelihood of illness and shorter life expectation than their Japanese equivalents, and every aspect of lifestyle has been examined to discover why.

Diet is a major factor in everyone's health, and the greater importance in the Japanese diet of fish, especially oily fish, fresh fruit and vegetables, and the reduced reliance on processed foods has been thought to be of significance. Algae are part of many people's staple diets, a popular dish in Japan is seaweed – including such varieties known as Kombu, Wakame and Nori – and the use of dried green and blue-green algae as food supplements started in the Far East, where some dried microalgae command a very high price. It is, then, hardly surprising that algae and microalgae have attracted the interest of health writers and consumers in the west, first in the United States and then in Europe.

The quotation from *The Sunday Times* in the introduction to this book touches on some of the most important health concerns of our busy modern lives. Many people buying this book will be doing so because they have read or been told that blue-green algae will 'boost vitality ... improve sleep, reduce allergies, stop migraines, reduce stress, alleviate PMT and boost the body's immune system'. The success of green and blue-green algae in recent years can indeed be attributed to the remarkable properties they appear to have to help with these conditions.

Sleep

Herbalists have for hundreds of years been aware of the help that valerian, lettuce and other plants and herbs can give to people who have difficulty sleeping, and it may well be that such properties are shared by blue-green algae, and Spirulina in particular. A balanced diet containing all the vital minerals and vitamins is essential to your body's well-being, and a healthy person is usually able to sleep without difficulty. Stress often interrupts sleep patterns, and if you start taking more care of your health you will learn to avoid stress, or to balance periods of stress with time for reflection or meditation. A diet containing blue-green algae can ensure that you are regularly taking in many of the vitamins, minerals and trace elements that your body needs to stay healthy. Moreover, many people report that they cope far better with periods of stress, and consequently enjoy more regular sleep patterns, once they have started on a course of green or blue-green algae.

Allergies

Allergies can seriously affect your quality of life, and it is important to take steps to eliminate anything from your life that may produce an allergic reaction. Strategies to identify allergens by systematically omitting common foodstuffs such as chocolate, milk products, caffeine and so on can unfortunately be accompanied by vitamin and mineral deficiencies which a course of green or blue-green algae can remedy. Algae are generally free of allergens, and indeed many people say that including them in their diet can counter allergic reactions to foodstuffs, chemicals or other materials.

If you are prone to allergic reactions, then it is recommended that you use dried green or blue-green algae in loose form rather than in capsules.

Migraine

Only someone who has experienced the debilitating symptoms of migraine can truly appreciate just how serious this condition can be. Despite considerable research, the causes of migraine are still little understood, and many remedies are taken more from desperation than from any scientific conviction of their effectiveness.

Sufferers who regularly include green and blue-green algae in their diets have said that their migraine attacks occur with less frequency; and research is under way to identify any biochemical explanation for this.

Pre-menstrual Syndrome

Women taking blue-green algae as a health food supplement report a reduction both in the severity of pre-menstrual symptoms and in the number of days they are affected by those symptoms. Try including algae in your diet for six months and note down any changes in the way you feel. Often it can take several months for the full effects to become apparent.

Degenerative Diseases

Many blue-green algae, including Chlorella, Aphanizomenon flos-aquae and Spirulina, are rich sources of gamma-linolenic acid which has been shown to have some importance in a number of degenerative diseases. Gamma-linolenic acid converts in the body to prostaglandin PGE_1, which is involved in numerous essential tasks in the body including the regulation of cell proliferation, blood pressure and inflammation of tissue. Gamma-linolenic acid deficiency is being investigated as a contributory factor in, amongst other conditions, schizophrenia, alcoholism and manic-depression. Gamma-linolenic acid is rare in ordinary foods but relatively abundant in Spirulina, which can yield the fatty acid without destroying the protein.

Disease Prevention

It is becoming increasingly clear that dietary minerals and vitamins can have a chemopreventive action on a variety of diseases, that is that certain substances in the diet can actively prevent illness and disease developing. The most extensive chemopreventive work is currently in the field of cancer.

The idea of using specific foods and supplementary nutrients to prevent disease, including cardiovascular disease and cancer, dates back to the early 1970s when increasing knowledge about environmental carcinogens raised researchers' hopes of also finding agents capable of protecting and stabilizing cellular and tissue differentiation against mutagenic agents such as free radicals. Free radicals are atoms or molecules that have one of their electrons missing. Electrons should be paired and the destabilizing effects of one electron trying to pair up with an electron from another molecule has the potential to cause damage to cells. Free radicals are the by-products of metabolism and are also in the toxins that can invade the body from such environmental factors as the smoke from cigarettes.

Antioxidants

Antioxidants are the body's natural defence against free radicals. They can repair missing electrons and so neutralize the free radicals, restabilizing the molecules before they damage other body cells. A number of vitamins, including vitamins C, D, E and beta-carotene, are thought to be antioxidants.

Green and blue-green algae are rich in antioxidant vitamins, so including them in your diet may well help you fight off disease by boosting your immune system. Trials suggest that antioxidants may have protective effects against diseases, and a lot of work in this area is concentrating on vitamin D. Individuals with low intakes of vitamin D, or with low serum levels of the vitamin, have been noted to be at an increased risk of colon cancer. The hypothesis that antioxidant vitamins might reduce cancer risk is based on a large body of both basic and human epidemiologic research.

A great many studies provide remarkably consistent data suggesting that consumption of foods rich in antioxidant vitamins reduces the risk of developing cancers. In the United States a number of randomized trials are currently under way designed to test the hypothesis that antioxidants prevent chronic diseases, and to evaluate the long-term safety of the widespread practice of supplementation.

Cancer

Epidemiologists have started suggesting that some environments and diet micronutrients might play a protective role against cancer. One study in Italy suggests a protective effect of vitamins A, C and E. Research has suggested that beta-carotene is effective in lung cancer and prostate cancer chemoprevention, but other trials have concentrated on alpha-carotene, vitamin A and its derivatives, vitamin C, selenium compounds, combinations of calcium with vitamin D_3, vitamin D_3 analogues, and vitamin E. There is strong and consistent evidence that intakes of fruits and vegetables protect against a variety of forms of cancer, making vitamins and minerals commonly found in the diet prime targets for chemoprevention studies. The supplementation of the diet of 30,000 Chinese residents of Ling Xian County with vitamin E, beta-carotene and selenium led to a reduction, after five years of use, in the incidence of and mortality from all cancers, and in particular cancer of the stomach.

Oral supplementation with calcium (1.5–2g per day) is currently being evaluated in a number of studies in the United States (in combination with vitamins and other antioxidants in one study, and with dietary fibre in another). Extracts from Spirulina and Dunaliella, fed by mouth to hamsters, was shown to be a powerful chemopreventive of oral cancer. In the Czech Republic freeze-dried blue-green algae was found to reduce mutagenic activity. Animals fed with beta-carotene showed a reduction in tumour number and size, but those fed with the blue-green algae showed a complete absence of tumours. Aqueous, ethanolic and chloroformic extracts of some 13 algae from both the Arabian Gulf and the North Adriatic Sea exhibited activity inhibiting the growth of human tumour established cells.

Dunaliella, which is rich in beta-carotene, has been shown to inhibit mammary tumours in mice, and to promote the growth of normal mammary gland cells and to inhibit neoplastic cells by normalizing both the specific organ and the general metabolism. It is likely that in the future the use of chemopreventive drugs will be the norm for people who have had primary cancers or who are considered for different reasons to be in a high risk group (people with a family history of cancer, smokers, etc.). The role of blue-green algae is proven of importance is this area.

HIV

The properties of the world's algae are slowly and painstakingly being investigated by the pharmaceutical industry. Many algae, especially seaweed, have been traditionally used for their healing properties and it is these that are the first targets for pharmaceutical exploitation. Since HIV has become widespread there has been a concentrated investigation into the properties of plants in the search for anti-virals. Extracts of several algae, including Asterionella notata, have been found to have anti-viral and anti-fungal properties. The antibiotic acrylic acid is commonly found in other microalgae, as are phenols, which are anti-microbials.

In 1989 Dr Michael R. Boyd, working at the National Cancer Institute in the United States, announced what was described as 'very preliminary findings' that chemicals derived from blue-green algae were found to be 'remarkably active' against the AIDS virus. Writing in the *Journal of the*

National Cancer Institute, Dr Boyd and his colleagues at the National Cancer Institute reported that they had produced cellular extracts from a series of blue-green algae, including Lyngbya lagerheimii and Phomidium tenue, to screen them for anti-human immunodeficiency virus (HIV–1) properties. It was found that at non-toxic concentrations the extracts were strikingly active against HIV–1 in cultured human cells. The blue-green algae were gathered from Hawaii and the Palau Islands, where they are prolific, and grown in the laboratory before being freeze dried prior to preparation of the extracts. A study of over 900 strains of blue-green algae published in 1993 found stronger evidence that extracts of certain cyanobacteria have the ability to inhibit the reverse transcriptases (transformers of normal cells to tumour cells) of avian myelobastosis virus (AMV) and HIV–1.

Appetite Suppressant

The addition of Spirulina to the feed of chickens has been shown to reduce their food intake overall, and in consequence to retard growth. If you take a small amount of Spirulina (1–3g) before a meal, then your appetite is reduced. Because of this, Spirulina is sometimes marketed as a diet pill.

Fertility Enhancement

Beta-carotene-rich Dunaliella fed as a supplement has been conclusively demonstrated to enhance body growth in male mice and to increase litter size and rearing rate. Because of this, Dunaliella is being investigated for the enhancement of fertility in humans.

Natural Deodorant

Chlorophyll from Spirulina, which contains iron oxide and a higher oxide, has been patented for its deodorant properties. It is already being used in products labelled as 'natural' deodorants.

Bronchitis

Algasol – an algal phycocolloid developed in Italy and reported as having 'exerted a useful oncological effect' (in other words, as having countered a number of cancers) has also been used with patients suffering from chronic bronchitis or emphysema.

Wound-healing Agent

Alcoholic extracts of Scenedesmus have been tested in the Czech Republic as constituents of ointments both for eczema and for ulcers, burns and wounds that have shown a reluctance to heal. Out of 109 patients observed, an astounding 91 per cent were healed, with a further 7 per cent showing some improvement. There were 112 control patients of whom only one person showed any improvement. Not surprisingly, the Czech Republic has funded these findings and has looked at more microalgae, with the best results being shown by children with eczema. In Africa poultices containing extracts of Spirulina have traditionally been found to hasten the healing of wounds. Under research conditions it has been discovered that enzymatic hydrolyzates of Spirulina promote skin metabolism and prevent keratinization. Pharmaceutical companies are following up these findings by producing cultures of Spirulina and adding them to creams, ointments and suspensions.

Blood-clotting

Some algae have significant properties that prevent the blood from coagulating. Over 60 species – the most researched being agar, iridophycan and carrageenan, which are derived from red seaweed, and fucoidan, derived from brown algae – are reported to have such properties. Scenedesmus has been similarly used in post-operative care.

Duodenal ulcers

Chlorella tablets have been used to great effect in Japan in the alleviation of gastritis and gastric and duodenal ulcers. Taking the microalgae by mouth also helped in the healing of 'incurable' wounds. Derivatives of Irish Moss, a red seaweed, have been used in the UK to strengthen the lining of the stomach against the damaging effects of ulcers.

Herpes

A number of polysaccharides from red seaweed have been shown to have anti-herpes properties. Extracts of Pacific Rhodophyceae were shown to be 99 per cent effective in stopping viral multiplication, and 50 per cent effective in reducing the spread of existing infections.

Laxative

Agar, derived from red seaweed, is a natural laxative which has been used for centuries. Derivatives of agars have long been used as binding agents for tablets, and alginates, which are digested in the gut rather than the stomach, provide a valuable suspensory agent for drug capsules. Other agars are used to produce plates on which bacterial and fungal cultures are grown.

Tapeworm

In Japan one red seaweed is the active ingredient in a popular preparation against tapeworms.

Bladder Disorders

Seaweed, traditionally used in the treatment of bladder disorders, are currently being analysed for their therapeutic compounds.

Coronary Atherosclerosis

Seaweed are recommended as a supplement to an anti-atherosclerotic diet and in the treatment of and prevention of coronary atherosclerosis.

Cholesterol-lowering Agent

It has been demonstrated that certain microalgae – in particular Spirulina and Scenedesmus – possess cholesterol-lowering properties. Both blood cholesterol and depositions of cholesterol in the liver were dramatically reduced in cholesterol-stressed animals fed Scenedesmus.

Anti-tumour Agent

The Japanese are using an extract from the seaweed Wakame in drug protocols to investigate their use as an anti-tumour agent, and there have been interesting results from tests on animals.

Morning-after Pill

The Sri Lankan marine red algae Gracilaria corticata, Gelidiella acerosa and Jania have been found to have post-coital contraceptive properties in rats. A similar preparation for use with humans would require considerable testing for toxicity, etc., but the potential is clearly there.

UV Screen

A pigment called scytonemin in certain blue-green algae has been found to provide significant protection against damage from ultraviolet radiation. The research is of importance to our understanding of plant ecology.

Lung Disease

Derivatives from microalgae are being investigated in the treatment of lung disease and scrofula.

Goitre

Research is currently being undertaken to exploit the toxins that are common to many algae. Kelp is rich in iodine and so is a natural prophylactic against goitre – enlargement of the thyroid – said to explain the low number of cases of goitre in countries in the Far East in which seaweed forms a major part of the diet. A Russian study, on the other hand, encouraged the use of Spirulina as an animal feed because the abundance of iodine in Spirulina was observed to stimulate growth in animals fed on the microalgae.

5

What Else Can It Do?

Scientists working on blue-green algae in Italy, and in particular on the potential of Spirulina, have said that the role that
blue-green algae may play in various important aspects of
life and human society depends to a great extent on the inge-
nuity and endeavours of the scientific community. While
many thousands of species of microalgae have been named,
only the smallest fraction of those have been examined in
terms of their potential for commercial exploitation, and it
has been said that it might well be possible to produce from
microalgae all the products that are at present derived from
other agriculture.

In conservation terms, microalgae are extremely efficient.
Being unicellular they contain all their nutrients and chemicals in their entire biomass – there are no roots, leaves or
fruit. They grow abundantly, are not seasonal and can thrive
in the harshest of conditions – conditions that would not
support conventional agriculture.

Some algae, mostly the red and brown seaweeds
Rhodophyta (Gracilaria, Porphyra, Rhodymenia) and Dulse
(Rhodymenia), are collected and used directly in the local
diet, the first by Polynesians and the latter by North Ameri-
cans. Spirulina was supposedly eaten by the Aztec civiliza-
tion in Mexico and still forms an important part of the diet of

the Kanembu people of Western Africa who make it into
dried cakes called *dihe*. Other algae are exploited indirectly as
food. Red algae (agar agar, for example) are used for their
gelling agent properties in commercial foodstuffs, such as jel-
lies and ice-creams.

There are now many chemicals that are being extracted
from commercially cultivated microalgae. Spirulina is being
used to produce phycocyanin and linoleic acid; Dunaliella for
glycerol and beta-carotene; Porphyridium for polysaccharides and arachidonic acid. Phycocyanin, as a natural pigment, is being marketed by Dainippon Ink & Chemicals of
Japan as a pigment for the food, drug and cosmetics industries under the brand name Linablue. It has been described
as 'an odourless non-toxic blue powder with a slight sweet-
ness, and has brilliant blue with a faint red fluorescence in
water'. It has been promoted as ideal for eye shadow, eye
liner and lipstick.

Animal Foodstuff

Microalgae have an important role indirectly for the human
chain since they are used in aquaculture and as foodstuff for
farm animals. The high concentrations of carotenoids has
produced much-sought-after enhancements in the coloration of meats and egg-yolks. Provided that the microalgae
fed to hens does not have a strong taste it is proving to be an
ideal source of nutrients and produces a first class product.
The premium put on the price of eggs produced by hens liv-
ing in ideal and 'humane' conditions does not seem to have
deterred the consumer who demands excellence of taste and
appearance as well. Foodstuffs from microalgae are not
cheap but many people will pay extra to avoid the fish smell
of factory-farmed eggs.

The use of microalgae in the diets of farmed fish has also

proved successful. One study used a green alga rich in carotenoids in the feeding programme of rainbow trout, providing what was described as 'visual enhancement of flesh coloration'.

Water Purification

There are many spin-offs from blue-green algae, both naturally occurring and under cultivation. Spirulina, for example, has been encouraged as a natural water-purifying factor in tropical and sub-tropical countries. Chlorella grows particularly abundantly on water contaminated by sewage and in hot conditions. Work at the University of Baroda in India in the 1960s sought to discover the nature of the algal-bacteria symbiosis with different species of green and blue-green microalgae and diatoms in waste treatment systems. Both Chlorella vulgaris and Oscillatoria chalybea were found to inhibit the metabolic activity of bacteria, thus providing a degree of purification. In warmer countries the establishment of algal ponds is now the most economic method of treating liquid agricultural waste and sewage. An oil-degrading bacterium associated with blue-green algae has been found in oil-polluted soil. The micro-organisms biodegraded 50 per cent of oil spills on the Arabian Gulf coast within 10 to 20 weeks. One American patent has been registered that uses a polysaccharide from a microalga to recover petroleum from crude oil. The culture outperformed a leading commercial biopolymer. Other research in the US has shown that some cyanobacteria have a natural ability to degrade pesticides.

One of the spin-offs of the use of microalgae in sewage treatment is the production of methane. Anaerobic digestion which results in the production of the gas is an artificial system which copies the natural biological process for the recycling of organic materials. As well as the gas, the resultant residue is a rich and valuable fertilizer. Microalgae are used in waste water treatment where blooms of algae can produce large quantities of oxygen. Scenedesmus obliquus was used in experiments in England with domestic sewage ponds to see whether the microalgae would aid in the removal of faecal contaminants. E. coli was removed within four days. Other microalgae investigated include Chlorella and Spirulina which were found to remove both nitrogen and phosphorous content. In Japan the technology of using micro-organisms in the treatment of waste, including both industrial waste and domestic sewage, is well advanced. Microalgae have been shown to be successful in removing specific substances such as phosphorus, nitrates and heavy metals. Even such environmental pollutants as lead, mercury and cadmium have been removed by microalgae which concentrate the metals and transform them into less hazardous forms.

Nitrogen-fixing

In the commercial field the possibility of biophotolytic hydrogen production has been investigated by researchers in Germany and Japan. Some species of algae generate hydrogen under incubation conditions and it is feasible to use such a method to produce hydrogen, but the economics of such a process rule out the possibility in the near future. Blue-green algae are the only organisms that combine the metabolic capabilities of photosynthetic oxygen production and nitrogen fixation. In south-east Asia algae grow in combination with rice in paddy fields and act as a nitrogen-fixing agent. In India a national project has taken this phenomenon and co-ordinated an investigation into the application of algae as biofertilizer.

Chemicals for Rubber Production

Kelp has been used for centuries and some kelps yield vital chemicals used in the production of rubber goods, including tractor tyres.

Paper

Blue-green algae derivatives are used as industrial lacquers to strengthen paper.

Industry

Red algae, including Chondrus and Gelidum, have a surprising variety of industrial uses, for example in the drinks industry, in the preparation of cosmetics, as a coating for photographic film, in the production of paints and in the meat industry. Dental impressions make use of red algae and many thousands of kilos are harvested world-wide for this purpose annually, although this is fast becoming replaced by synthetic products.

Colorants

The red microalgae contain pigments called phycobili-proteins, which are red or blue, and which are used as colorants in food processing, cosmetics and pharmaceuticals, in particular as a 'natural' substitute for synthetic dyestuffs.

Cosmetics

Irish Moss (Chondrus crispus) and other red seaweed provide carrageenan, a phyco-colloid, and derivatives are used

in toothpaste. Combined with potassium it provides raw
materials for hair preparations.

Much has been written about the potential use of blue-
green algae in space stations, and the word 'algatron' was
coined for a system that employed algae to produce oxygen
to maintain human life. Humans breathe in oxygen and
breathe out carbon dioxide. Plants convert carbon dioxide
into oxygen. A person requires 600 litres of oxygen every 24
hours and it has been estimated that if enough of the blue-
green alga Chlorella was growing in the environment of the
space station to produce 600 litres of oxygen it would in turn
consume 700 litres of carbon dioxide. The algae would in the
process grow by an estimated half kilo (dried weight). Since
Chlorella is almost 60 per cent protein, 10 per cent fat, 20
per cent carbohydrate and 3 per cent fibre it would appear to
be the perfect symbiosis. In the late 1980s Chlorella was
flown as a component of an algobacterial cenosis-fish system
during the Russian space flight Cosmos 1887. No differences
were observed between the growth rate or development of
the blue-green algae on board the spacecraft and on land.

6

Microalgae You are Most Likely to Hear About

Chlorella

Chlorella are tiny spherical non-motile unicellular green algae which reproduce asexually – as cells mature they divide. Virtually all the world's commercial output of Chlorella comes from Taiwan, where production started in 1964 and where the number of factories has grown from 30 in 1977 to over 100 today, producing many thousands of tons a year for the health food market. It is available in loose dried form, in tablets and in capsules. A product called Chlorella Growth Factor accounts for a large part of the multi-million dollar market for the product. Chlorella are increasingly appearing in health food preparations including cakes and drinks. The economic potential of the polysaccharide content of Chlorella is being exploited commercially. The starch content of one particular strain of Chlorella can reach approximately 50 per cent of the total dry weight.

Dunaliella

Dunaliella are ovoid unicellular green algae. The cells are motile since they have two long flagella and they contain one large chloroplast which takes up about half the volume of the cell. They reproduce asexually and seem to like brackish or salt water conditions. They thrive in the Dead Sea, the land-locked salt lake between Israel and Jordan, and at the Great Salt Lake near Salt Lake City in the United States. Microalgae naturally adjust to extremes of salinity by a process called osmoregulation with the consequence that they concentrate production of valuable chemicals.

Koor Industries in Israel are exploiting the ability of Dunaliella to adapt to the saturated NaCl (sodium chloride or common salt) conditions to develop large-scale production of glycerol and beta-carotene. Glycerol is an alcohol, used com-mercially as glycerine or in the form of glycerides in a vast number of industries from paints to foods, cosmetics and pharmaceuticals. Under ideal conditions Dunaliella can yield 16g of glycerol per square metre per day. Beta-carotene is used as a source of provitamin A and as a natural food colouring. Beta-carotene from Dunaliella is marketed as 'natural' or 'organic' and so commands a higher price than synthetic beta-carotene.

Scenedesmus

Scenedesmus are multicellular green algae which reproduce both sexually and asexually. For habitat they generally prefer fresh water or soil. They are rich in protein, vitamins and minerals but their chlorophyll levels are high, and so they are highly coloured, and they have a strong taste. There have been attempts to introduce Scenedesmus into the range of microalgae offered as food supplements, but with little suc-cess. However, they have been used extensively in animal feed and there are attempts to introduce the microalgae com-mercially as a food-extender.

Aphanizomenon flos-aquae

Hollywood's favourite, Aphanizomenon flos-aquae is a multicellular cyanobacterium which has been the subject of considerable interest over the last decade. It grows abundantly on lakes and is found predominantly in the northern hemisphere. It gathers in prodigious amounts and has the appearance of tiny filamentous blades of common grass. It famously thrives at the breath-taking Klamath Lake in Oregon State in the United States which is fed by a series of icy mountain streams. Unlike most of the other commercially available microalgae and cyanobacteria, which are cultivated, most of the Aphanizomenon flos-aquae marketed in Europe and the United States is harvested directly from the lake and has provided a growing business for the area. Its nutrients and chemical composition make it an important candidate for future research – processing renders it as digestible as other microalgae, if not more so, and it retains more of its nutritional properties in the drying process.

However, like a number of other cyanobacteria, Aphanizomenon flos-aquae is known occasionally to produce neurotoxins. The most commonly reported toxin is the same one responsible for paralytic shellfish poisoning which can affect fish stock, livestock and humans. Exactly why some cyanobacteria are found to be toxic is not entirely understood, although it is thought that the conditions, including the pH of the medium in which the blue-green algae are grown, bear some relation to the existence of toxicity. None of the Aphanizomenon flos-aquae from Klamath Lake has been found to be affected and many more microalgae that grow naturally are non-toxic. Of course, all blue-green algae that are sold over the counter are carefully screened for toxins and contaminants.

Spirulina

Spirulina are multicellular, filamentous cyanobacteria which are exceedingly adaptable and occur in a wide variety of environments including seawater, fresh water, tropical springs and salt pans. Under the microscope Spirulina is blue-green in colour and, as its name suggests, has the appearance of a spiral of long thin threads. Spirulina are relatively large, ranging between 1µm and 12µm in diameter and between 3mm and 20mm in length. Famously, Spirulina is harvested from predominantly alkaline lakes in Central Africa, where it is dried in the sun and eaten as a cake. Spirulina grows prodigiously, colonizing some very extreme environments, and is one of the most abundant algae in the salty alkaline lakes in Africa and in North and South America. It is full of nutrients and easily digested. It is Spirulina that caused so much interest in France after a Belgian military expedition crossing the Sahara from the Atlantic to the Red Sea in the 1960s noticed that the Kanembu people made cakes from the blue-green algae that appears to choke Lake Chad.

Today Spirulina has penetrated the health food market successfully and is commercially available as powder and in tablet and in capsule form, but it has also made the transition into conventional processed food, being used in at least one soy-whole-wheat noodle product. The American Federal Drug Administration (FDA) ruling reads: 'Spirulina is a source of protein and contains various vitamins and minerals. It may be legally marketed as a food or food supplement so long as it is labeled accurately and contains no contaminated or adulterated substances.'

The annual world production of Spirulina is estimated at 500 tons, with commercial production taking place in Israel, Japan, South America, Taiwan, Thailand and the United

States. In Mexico, at Lake Texcoc, Spirulina grows naturally and a health food company is harvesting the biomass direct from the Lake.

Less-well-known Species

There are other less-well-known species under mass culture. Phaeodactylum has potential for the commercial production of lipids, Botryococcus for hydrocarbons – possibly becoming a source of renewable liquid fuel. Porphyridium is important for polysaccharides which are valued for their gelling qualities and for a derivative which has been patented for use in the recovery of oil. Porphyridium is also valued for use as a pigment and as a source of arachidonic acid – the subject of considerable ongoing research. Porphyridium is one of the many algae to be valued as a foodstuff.

7

How It's Grown

Most microalgae are so small that the processes used to harvest them successfully can be technologically intensive, time-consuming and consequently expensive. They can be cultivated under natural conditions – in situations where they accumulate on the surface of lakes – but for various reasons, not the least of which is the potential for microalgae to develop or accumulate heavy metals and toxins, most blue-green algae in commercial use are cultivated in huge artificial ponds.

The key component in the production of algae is, of course, a strong and reliable source of sunlight. However, the amount of movement displayed by the algae, different intensities of light, and varying temperatures can sometimes have photoinhibitory effects on algae affecting photosynthetic response and chemical composition. Although it may seem that since microalgae grow abundantly in nature that little husbandry might be required in the mass culture of algae, the opposite is in fact the case. Constant measurement of temperature, oxygen levels, population density and pest levels is essential. Contamination by bacteria, fungi, viruses and zooplankton is inevitable. These parasites can take up residence and completely destroy some cultures. Chlorella, Scenedesmus and Aphanizomenon flos-aquae are particularly prone to depletion by parasites, although Spirulina has proven hardier in so far as both pests and viruses are concerned.

Many countries around the world have developed commercial microalgae cultivation, including Bulgaria, the Czech Republic, Japan, Taiwan, Thailand and the United States, and techniques vary from country to country. Three aspects of the mass cultivation of algae affect set-up costs, ongoing costs, yield and value of the finished product. The first is the size and shape of the 'reactor systems' in which the algae are cultivated. The second is how to separate the algal mass from the medium in which it flourishes. The third is the method adopted to dehydrate the algal mass for storage and distribution.

Two designs are currently used in various parts of the world. In one, huge oblong troughs are used in which the water is constantly agitated by paddle wheels. If the troughs are covered (they are sometimes long polyethylene tubes) then temperature can be more easily controlled while saving the medium from evaporation. This is not without a corresponding loss of exposure to sunlight from water condensation, accumulation of dust and debris, etc.

A second reactor design, developed in eastern Europe, uses a gently sloping 300 square metre concrete pond containing a culture depth of some 20cm in which the medium is recycled by pumps. A joint Peruvian-German algal plant at Sausal in Peru has produced a yield of over 50g per square metre per day (Scenedesmus and Chlorella) employing this system.

In terms of cost, the construction of the ponds, and in particular the cost of replacing the linings (under constant attack from UV and water), is all important. Moreover, the algae need to be in a process of constant agitation to prevent them from sinking, with consequent deterioration in overall penetration of sunlight and build-up of dissolved oxygen.

Carbon dioxide is required, as well, and there are differing methods of delivering a supply so that as little of the gas as possible is lost to the atmosphere. It is the nutrients that are required for the mass cultivation of microalgae that add significantly to their cost. As well as carbon dioxide they need nitrogen. If the culture is sufficiently alkaline, then the carbon nutrient is reduced. As far as nitrogen is concerned, if nitrogen-fixing organisms could be cultured in conjunction or in symbiosis with the algal species, then the requirement for nitrogen fertilizer would be reduced.

The algae are collected suspended in what can only be described as slurry, and they need to be extracted from this medium so that the filtrate can be recycled. Centrifugation is one method of reducing the water content, others include sand-bed filtration and techniques such as chemical flocculation and electroflotation. Centrifugation has the advantage that no added chemicals are used but the machinery and energy resources required add to the cost of collecting the algae. Cyanobacteria – being filamentous – can be filtered with screens and strainers. Spirulina, for example, is harvested using vibrating screens. Chemicals added to the algal mass can help the cells to adhere together (floc) or in certain conditions this may be spontaneous. Flocculation can also be induced by electricity.

The microalgae are then further prepared by sun drying, spray drying or freeze drying. Each method has disadvantages, whether it be in terms of damage done to the nutritional content, taste, digestibility or, again, the relative final cost – the drying process accounting for up to 30 per cent of the cost of production.

Each method affects in different ways the food values and taste. Research has shown that when Spirulina is spray dried it loses between 7 and 10 per cent of its beta-carotene. In coloured bottles containing the slightest amount of air, more than 50 per cent is lost after 45 days. Even the size of the dried material makes a difference: flakes retain more of the beta-carotene than fine powder. In work done in Taiwan, Chlorella and Spirulina were spray-dried and freeze-dried under varying conditions and the morphological changes induced were studied by scanning electron microscope. The internal structure was also examined using cryofracturing. The methods of drying, the time taken, the temperature and the processes involved had effects on the external morphology of the powders.

Spray-drying and freeze-drying are the most expensive techniques and are the favoured processes for the health food industry – both producing an attractive powder. Drum-dried green algae are said to be more easily digested. Spray-drying is the technique used to produce powdered milk – a solution or suspension is sprayed into a flow of hot gasses and so rapidly drying to form a powder. Surface drying uses a rotating steam-heated drum on which the algal solution is spread. Contact with the surface of the drum is for a few seconds only but at a high enough temperature (120°C) to cause dehydration. This also ruptures the cell wall. Freeze-drying is the evaporation to dryness of the algal solution and the process maintains most of the nutrients and pigments.

Untreated algae are indigestible to anything other than cattle, sheep and goats because they have a thick cellulose-containing cell wall. These animals produce an enzyme called cellulase which hydrolyses the cellulose cell wall and so makes the protein digestible. For algae to be available as human food the algal cell has to be opened by one of the drying methods or by boiling. Of the common blue-green algae, only Spirulina's cell wall is comparatively thin, but even Spirulina requires some processing. This is also why perhaps the most attractive method of drying is inappropriate for microalgae – evaporation in the open air or sun-drying.

Given the energy needs and the overall abundance of sunlight in microalgae cultivation locations it would seem that costs could be reduced by making use of solar energy for a number of the processes here described. There is too much potential in microalgae for production to be confined to the health supplement market simply because of cost.

8

The Down Side

As has been seen, microalgae can thrive in many climatic extremes and have a tendency to proliferate. Consequently, some microalgae can overproduce, or 'bloom', ruining fresh water sites used for drinking water or for recreational purposes. Under these conditions the oxygen in the water can be used up, leading to the complete greening of the surface of the water and the subsequent loss of fish stocks. In this way microalgae can threaten the whole ecological balance of particular environments. Fresh water can become tainted in taste and unpleasant smells can emanate from localities where microalgae are over-abundant.

Some writers maintain that blue-green algae can cause serious public health problems. In tropical areas, dinoflagellates (Gymnodinium) in rivers have been known to cause the death of fish. In some areas where shell fish have fed on the same algae, toxins can accumulate in the shell fish themselves and poison humans who eat them. Unfortunately, toxic strains of some species of cyanobacteria are indistinguishable morphologically from non-toxic strains. The toxic algae have caused fatalities when fed to livestock and are known to cause gastro-enteritis in humans.

Algal toxins can affect the drinking water supply. There are numerous reports of farm animals, wildlife and even domestic animals being poisoned by microalgae, and there are some cases of humans suffering from the effects of the algal blooms. Contamination of sea and freshwater fish with toxins has been reported from Australia, and reports of the poisoning of cattle and wildlife from drinking water supplies have been increasing during the last few years throughout the country. In the 1960s outbreaks of gastro-enteritis among children who lived near Lake Chivero in Zimbabwe were linked to cyanobacterium Microcystis aeruginosa.

Toxic blue-green algae may be ingested from drinking water, but the toxins can also be released into water if the microalgae are killed – by, for example, the chlorination of a swimming pool. A reservoir may thus be contaminated first by toxic algae and then by the free toxins – which neither conventional water filtration nor boiling will remove. Although ill-effects are generally confined to livestock and domestic animals there are clinical reports of gastro-intestinal disorders in people who have ingested cyano-bacteria whilst swimming. Symptoms include headache, stomach cramps, nausea and painful diarrhoea. Dermatitis is sometimes a problem with cyanobacteria, and allergies have also been reported, including severe skin reactions, hay fever, asthma and eye irritation. In 1977 one batch of Chlorella tablets caused rather nasty skin lesions in some of the people who had taken them, and it was later discovered that photoconversion of chlorophyll to pheophorbide had taken place.

Microalgae have the potential to bioaccumulate heavy metals, and it is possible that this could lead to a successive accumulation of those heavy metals in other parts of the food chain, with consequent serious threats to the health of humans either eating microalgae directly or eating animals fed on microalgae.

The cyanobacterium with most concerns in this area is

Aphanizomenon flos-aquae. To date no problems have been reported from the Aphanizomenon flos-aquae taken from the Klamath Lake area in Oregon. The cyanobacterium is marketed as being 'organic' and is one of the few commercially exploited natural sources of microalgae – most food-grade microalgae being cultured under strict conditions in factories run as laboratories. Aphanizomenon flos-aquae is, however, one blue-green alga known under certain conditions to produce toxic blooms.

Of course, all blue-green algae available commercially are tested for toxins and no case has ever been reported of poisoning from mass cultivated algae, where the conditions are closely monitored.

9

The Future

Relatively few of the many thousands of microalgae and cyanobacteria have been thoroughly researched for their economic, therapeutic and food-source potential. Moreover, since they are unicellular and have a short life-cycle, microalgae are ideal for genetic engineering. It should be possible to transfer the most useful characteristics of other plants to produce an organism that thrives in the available conditions and produces precisely the right mix of nutritional or chemical elements.

For livestock, as the scale of production increases the cost of the product will drop, so it is increasingly likely that microalgal foodstuffs will be able to compete in price with animal feeds that are at present in use.

The initial increase in algaculture was prompted by the health supplement and food market, with commercial production competing with 'natural' or 'organic' cultivation. Certainly the price of microalgae in health food stores ought to come down so that green and blue-green algae can become a part of everyone's diet and not just of those who can afford it. When we can purchase tubs of dried blue-green algae in the supermarket, in the same way that we can now obtain soya products, we will see more and more recipes appearing which make use of the extremely adaptable and unequivocally nutritious 'superfood'.

References

Aaronson, S., *Microalgae as Source of Chemicals and Natural Products: The Production and Use of Microalgal Biomass* (AKKO, Israel, 1978).

Anusuya Devi, M. and Venkataraman, L. V., 'Functional Properties of Protein Products of Mass Cultivated Blue-green Algae Spirulina platensis', *Journal of Food Science*, 49 (1984), 2 and 24.

Anusuya Devi, M., et al. (eds.), 'Studies on the Proteins of Mass-cultivated Blue-green Algae (Spirulina platensis)', *Journal of Agricultural and Food Chemistry*, 29 (1981), 522.

Barton, L. L., Foster, E.W. and Johnson, G. V., 'Viability Changes in Human Neutrophils and Monocytes Following Exposure to Toxin Extracted from Aphanizomenon flos-aquae', *Canadian Journal of Microbiology*, 26 (1980), 272.

Bayaanova, Y. I. and Trubachev, I. N., 'Comparative Evaluation of the Vitamin Composition of Some Unicellular Algae and Higher Plants Grown Under Artificial Conditions', *Applied Biochemical Microbiology*, 17 (1981), 292.

Becker, E. W., 'Algae Mass Cultivation: Production and Utilization', *Process Biochem.*, 16/5 (1981), 10.

Becker, E. W., 'Nutritional Properties of Microalgae:

Potentials and Constraints', in *Handbook of Microalgal Mass Culture* (CRC Press, Boca Raton, Florida, 1986), 339.

Bellinger, E. G., *A Key to Common Algae* (4th edn., London, 1992).

Benemann, J. R., Microalgae Biotechnology, *Trends in Biotechnology*, 5 (1987), 47.

Bewicke, D. and Potter, B.A., *Chlorella the Emerald Food* (Ronin Publishing, Berkeley, Calif., 1984).

Bold, Harold C. and Wynne, Michael J., *Introduction to the Algae: Structure and Reproduction* (1978).

Borowitzka, M. A., Large-scale Algal Culture Systems: The Next Generation, *Australasian Biotechnology*, 4 (1994), 212.

Borowitzka, M. A. and Borowitzka, L. J. (eds.), *Micro-Algal Biotechnology* (Cambridge University Press, Cambridge, 1988).

Borowitzka, L. J., 'Status of the Australian Algal Biotechnology Industry in 1990', *Australian Journal of Biotechnology*, 4 (1990), 239.

Boudene, C., Collas, E. and Jenkins, C., 'Study and Estimation of Some Mineral Poisons in Spirulina Algae of Different Origins and Long-term Toxicity Trial with Rats and Spirulina from Mexico', *Ann Nutr. l'Aliment.*, 29 (1975), 577.

Brock, T. D., Smith, D. W. and Madigan, M. T., *Biology of Microorganisms* (Prentice Hall, Englewood Cliffs, NJ, 1984), 760.

Carmichael, W. W. (ed.), *The Water Environment: Algal Toxins and Health* (Plenum Press, New York, 1981).

Carr, N. G. and Whitton, B. A., *The Biology of Blue-green Algae* (1973).

Cavi, G. de and Guiseppe, S., 'Bronchopathies, Complications and Outcome Under the Action of Algasol T–331', *Abstracts of the Eighth International*

Seaweed Symposium (Bangor, 1974).

Chapman, V. J., *Seaweeds and their Uses* (2nd edn., 1970).

Chapman, V. J. and Chapman, D. J., *The Algae* (1971).

Ciferri, O. and Tiboni, O., 'The Biochemistry and Industrial Potential of Spirulina', *Annual Review of Microbiology*, 39 (1985), 503.

Ciferri, O., 'Spirulina, the Edible Micro-organism', *Microbiological Reviews*, 47 (1983), 551.

Claudio, Virginia S. and Lagua, Rosalinda T., *Nutrition and Diet Therapy Dictionary* (3rd edn., Chapman & Hall, London, 1991).

Clement, G.,Rebeller, M. and Zarrouk, C., 'Wound-treating Medicaments Containing Algae', *Fr. Med.* (1967), 5279.

Collins, M., 'Algal Toxins', Microbiology Review, 42 (1978), 725.

Combs, W. et al., 'In Vivo and In Vitro Effects of Beta-Carotene and Algae Extracts in Murine Tumor Models', *Nutr. Cancer*, 12 (1989), 371.

Dagnelie, P. C., van Staveren, W. A. and van den Berg, H., 'Vitamin B_{12} from Algae Appears not to be Bioavailable', *American Journal of Clinical Nutrition*, 53 (1991), 695.

Dellweg, H. (ed.), *Biotechnology* (Verlag Chemie, Deerfield Beach, Florida, 1983).

Desikachary, T. V. (ed.), *Taxonomy and Biology of Blue-green Algae* (University of Madras, 1972).

Dunlop, J. M., 'Blooming Algae', *BMJ* 302 (1991), 671.

Falconer, I. A., *Algal Toxins in Seafood and Drinking Water* (Academic Press, London, 1993).

Fica, V., Olteanu, D. and Oprescu, S., 'Use of Spirulina as an Adjuvant Nutrient Factor in the Therapy of the Diseases Accompanying a Nutritional Deficiency', *Rev. Med. Int.* 36 (1984), 225.

Fink, H. and Herold, E., 'The Protein Value of Unicellular Green Algae and their Action in Preventing Liver Necrosis', *Zeitscher Physiol. Chem.*, 305 (1956), 182.

Finney, K. F., Pomeranz, Y. and Bruinsma, B. L., 'Use of Algae Dunaliella as a Protein Supplement in Bread', *Cereal Chemistry*, 61 (1984), 402.

Fox, R. D., *Algoculture*, Ph.D. thesis (1983).

Fryer, L. and Simmons, D., *Food Power From the Sea: The Seaweed Story* (Mason Charter, New York, 1977).

Guerin-Dumartrait, E. and Moyse, A., 'Biological Characteristics of Spirulina spp.', *Annales Nutr. l'Aliment.*, 29 (1975), 489.

Gustafson, K. et al., 'AIDS-Antiviral Sulfolipids from Cyanobacteria (Blue-green Algae)', *Journal of National Cancer Institute*, 81 (1989), 1254.

Hills, C. and Nakamura, H., *Food from Sunlight* (University of the Trees Press, Boulder Creek, Calif., 1978), 329.

Hunter, P. R., 'Cyanobacteria and Human Health', *J. Med. Microbiol.* 36 (1992), 301.

Ikawa, M. et al., 'Comparison of the Toxins of the Blue-green Algae Aphanizomenon flos-aquae with the Gonyaulax Tamarensis var. Excavata Toxins', *Toxicon*, 20 (1982), 747.

Jackim, E. and Gentile, J., 'Toxins of a Blue-green Alga: Similarity to Saxitoxin', *Science* 162 (1968), 915.

Johnson, H. W., 'The Biological and Economic Importance of Algae. 3. Edible Algae of Fresh and Brackish Waters', *Tuatara* 18 (1970), 19.

Johnson, P. and Shubert, L. E., 'Accumulation of Mercury and other Elements by Spirulina (Cyanophyceae)', *Nutr. Rep. Int.* 34 (1986), 1063.

Juettner, F., 'Mass Cultivation of Microalgae and Phototrophic Bacteria under Sterile Conditions', *Process Biochem.*, 17 (1982), 2.

Hoppe, H. A., Levring, T. and Tanaka, Y., *Marine Algae in Pharmaceutical Science* (Walter de Gruyter, Berlin, 1979).

Hoppe, H. A. and Levring, T. (eds.), *Marine Algae in Pharmaceutical Science*, Volume 2 (Walter de Gruyter, Berlin, 1982).

Kausner, A., 'Algaculture: Food for Thought', *Biology Technology*, 4 (1986), 947.

Kay, Robert A., 'Microalgae as Food and Supplement', *Critical Reviews in Food Science and Nutrition* 30 (1991), 555.

Kopteva, Z. P., 'Biosynthesis of Thiamine, Riboflavin and Vitamin B_{12} by Some Blue-green Algae', *Mikrobiol. Zh.*, 32 (1970), 429.

Kopteva, Z. P., 'Biosynthesis of Biotin, Pyridoxine, Nicotinic Acid and Pantothenic Acids by Some Blue-green Algae', *Mikrobiol. Zh.*, 32 (1970), 555.

Lembi, C. and Waaland, J. (eds.), *Algae and Human Affairs* (Cambridge University Press, Cambridge, 1988).

Levring, T., Hoppe, H. A. and Schmid, O.J., *Marine Algae* (Walter de Gruyter, Hamburg, 1969).

Mahmood, N. A. and Carmichael, W. W., 'Paralytic Shellfish Poisons Produced by the Freshwater Cyanobacterium Aphanizomenon flos-aqua NH–5', *Toxicon*, 24 (1986), 175.

Maranesi, M., et al., 'Nutritional Studies on Spirulina Maxima', *Acta Vitaminol. Enzymo.*, 6 (1985), 295.

Mateles, L. D. and Tannenbaum, J. E. (eds.), *Single-Cell Protein* (MIT Press, Cambridge, Mass., 1968).

McArdle, R. N., et al., 'Protein and Amino Acid Content of Selected Microalgae Species', *Lebensmittel Wissenschaft und Technologie*, 27 (1994), 249.

Meydani, M., 'Vitamin E', *Lancet* 345 (1995), 170.

Nagasawa, H. et al., 'Inhibition by Beta-Carotene Rich

Algae Dunaliella of Spontaneous Mammary Tumourigenesis in Mice', *Anticancer Research*, 9 (1989), 71.

Nakaya, N., Homma, Y. and Goto, Y., 'Cholesterol Lowering Effect of Spirulina', *Nutr. Rep. Int.* 37 (1988), 1329.

Narasimha, D. L. R. et al., 'Nutritional Quality of the Blue-green Algae, Spirulina Platensis', *Journal of the Science of Food Agriculture*, 33 (1982), 456.

Payer, H. D. et al., 'Major Results of the Thai-German Microalgae Project at Bangkok', *Ergebnisse Limnologie*, 11 (1980), 41.

Powell, R. C., Nevels, E. M. and McDowell, M. E., 'Algae Feeding in Humans', *Journal of Nutrition* 75 (1961), 7.

Prasomzup, U., 'Effect of Different Levels of Algal Diets on Rats', *Food*, 1 (1979), 294.

Rezanka, T. et al., 'Determination of Fatty Acids in Algae by Capillary Gas Chromatography-mass Spectrometry', *Journal of Chromatography*, 268 (1983), 71.

Richmond, A. E., *Handbook of Microalgal Mass Culture*, CRC Press, Boca Raton, Florida (1986).

Richmond, A. E., 'Microalgaculture', *Critical Reviews in Biotechnology*, 4 (1986), 369.

Rodriquez, L. M., 'Microalgae as Sources of Food and Chemical Products', *Grasas y Aceites*, 32 (1981), 245.

Sano, T. and Tanaka, Y., 'Effect of Dried, Powdered Chlorella Vulgaris on Experimental Atherosclerosis and Alimentary Hypercholesterolemia in Cholesterol-Fed Rabbits', *Artery*, 14 (1987), 76.

Sano, T. et al., 'Effect of Lipophilic Extract of Chlorella Vulgaris on Alimentary Hyperlipidemia in Cholesterol-Fed Rats', *Artery*, 15 (1988), 217.

Santillan, C., 'Mass Production of Spirulina', *Experientia*, 38 (1982), 40.

Sasner, J. J. Jnr et al., 'Studies on Aphantoxin from Aphanizomenon flos-aquae in New Hampshire', in *The Water Environment: Algal Toxins and Health* (Plenum Press, New York, 1981).

Sautier, C. and Tremoilieres, J., 'Food Value in Spirulina Algae in Humans', *J. Ann. Nutr. Aliment.*, 30 (1975), 517.

Schwartz, J. et al., 'Prevention of Experimental Oral Cancer by Extracts of Spirulina-Dunaliella Algae', *Nutr. Cancer*, 11 (1988), 127.

Shelef, G. and Soeder, C. J. (eds.), *Algae Biomass Production and Use* (Elsever/North Holland Biomedical Press, New York, 1980).

Shoptaugh, N. A. H., 'Fluorometric Studies on the Toxins of Gonyaulax Tamarensis and Aphanizomenon flos-aquae', *Dissertation Abstracts International* B 39 (1979), 5363.

Stadler, T. et al. (eds.), *Algal Technology* (Elsevier Applied Science, New York, 1988).

Switzer, L., *Spirulina: The Whole Food Revolution* (Bantam/Toronto, 1982).

Trainor, F. R., *Introductory Phycology* (John Wiley & Sons, London, 1978).

van den Berg, H., Dagnelie, P. C. and van Staveren, W. A., 'Vitamin B_{12} and Seaweed', Lancet, 30 January 1988, 242.

Vanossi, L., 'Chlorella as Food', *Industrie Alimentari*, 21 (1982), 118.

Venkataraman, L. V. et al., 'Investigation on the Toxicology and Safety of Algal Diets in Albino Rats', *Food Cosmetics Toxicology*, 18 (1980), 271.

Wise, D.L., 'Fuel Gas Production form Selected Biomass Via Anaerobic Fermentation', in *Biological Solar Energy Conversions* (Academic Press, New York, 1977), 411.

Index